U.S. Department
of Transportation

**National Highway
Traffic Safety
Administration**

www.nhtsa.gov

DOT HS 811 339

July 2010

The Effectiveness of ABS in Heavy Truck Tractors and Trailers

DISCLAIMER

1. Report No. DOT HS 811 339	2. Government Accession No.	3. Recipient's Catalog No.
4. Title and Subtitle The Effectiveness of ABS in Heavy Truck Tractors and Trailers		5. Report Date July 2010
		6. Performing Organization Code
7. Author(s) Kirk Allen, Ph.D.		8. Performing Organization Report No.
9. Performing Organization Name and Address Evaluation Division; National Center for Statistics and Analysis National Highway Traffic Safety Administration Washington, DC 20590		10. Work Unit No. (TRAIS)
		11. Contract or Grant No.
12. Sponsoring Agency Name and Address National Highway Traffic Safety Administration 1200 New Jersey Avenue SE. Washington, DC 20590		13. Type of Report and Period Covered NHTSA Technical Report
		14. Sponsoring Agency Code
15. Supplementary Notes		

16. Abstract

Federal Motor Vehicle Safety Standard No. 121 mandates antilock braking systems (ABS) on all new air-braked vehicles with a GVWR of 10,000 pounds or greater. ABS is required on tractors manufactured on or after March 1, 1997, and air-braked semi-trailers and single-unit trucks manufactured on or after March 1, 1998.

The primary findings of this report are the following:
- The best estimate of a reduction by ABS on the tractor unit in all levels of police-reported crashes for air-braked tractor-trailers is 3 percent. This is based on data from seven States and controls for the age of the tractor at the time of the crash. This represents a statistically significant 6-percent reduction in the crashes where ABS is assumed to be potentially influential, relative to a control group, of about the same number of crashes, where ABS is likely to be irrelevant.
- In fatal crashes, there is a non-significant 2-percent reduction in crash involvement, resulting from a 4-percent reduction in crashes where ABS should be potentially influential. The age of the tractor at the time of the crash is not important. Rather, external factors of urbanization, road speed, and ambient lighting are influential and are accounted for in the final estimate.
- Among the types of crashes that ABS influences, there is large reduction in jack-knives, off-road over-turns, and at-fault involvements in collisions with other vehicles (except front-to-rear collisions). Counteracting are an increase in the number of involvements of hitting animals, pedestrians, or bicycles and, only in fatal crashes, rear-ending lead vehicles in two-vehicle crashes.

17. Key Words NHTSA; NCSA; FARS; State Data System; ABS; heavy vehicles; tractors; trailers	18. Distribution Statement Document is available to the public from the National Technical Information Service www.ntis.gov		
19. Security Classif. (Of this report) Unclassified	20. Security Classif. (Of this page) Unclassified	21. No. of Pages 46	22. Price

Form DOT F 1700.7 (8-72)   Reproduction of completed page authorized

Executive Summary

This report presents a statistical analysis of crash data in order to determine the effectiveness of antilock brakes in tractor-trailer combination vehicles. Federal Motor Vehicle Safety Standard No. 121, Air Brake Systems, mandates antilock braking systems on virtually all new air-braked vehicles with a GVWR of 10,000 pounds or greater. ABS is required on tractors manufactured on or after March 1, 1997, and air-braked semi-trailers and single-unit trucks manufactured on or after March 1, 1998.

The first stage of the analysis considers ABS on both the tractor and the trailer. The amount of data is insufficient to draw conclusions about the effect of ABS on the trailer or whether it is beneficial for only the trailer to have ABS when the tractor does not. Next, other data sources are incorporated that identify the tractor but not the trailer. Data from seven States is used in a preliminary estimate of the effectiveness of ABS for the tractor in all levels of police-reported crashes. The median effectiveness from these seven States is 13 percent, in terms of reducing crashes where ABS could be influential compared to a control group (of about the same number of crashes) where ABS should be irrelevant. This represents a median reduction of 7 percent in terms of all crashes in the State Data System. The age of the tractor at the time of the crash is influential. Several methods of adjusting for tractor age suggest a slightly lower but still statistically significant effectiveness of 6 percent in terms of crashes where ABS should be influential (or 3% in terms of all State-reported crashes) – 6 percent is the median of estimates from seven States that range from 5 percent to 10 percent. Two factors may cause the observed effect to be an underestimate: (1) an unknown proportion of the tractors produced prior to the 1997 ABS mandate were voluntary equipped with ABS; and (2) an unknown proportion tractors produced after the 1997 mandate may have had ABS not functioning properly at the time of the crash.

The favorable result in all crashes is offset by a failure to find a statistically significant reduction in fatal crash involvements. Certain crash mechanisms are greatly reduced (e.g., jackknife) while others happen relatively more often for combination vehicles where the tractor is ABS-equipped (e.g., rear-ending a leading vehicle). It is plausible that the circumstances leading to a fatal crash are too overwhelming and multifaceted for ABS in itself to prevent the crash.

The effect on fatal crashes varies by the type of road. ABS significantly reduces crashes on non-interstate roads with speed limits less than 50 mph. On interstates and roads with speed limit 55 mph or higher, tractor-trailers rear-ending leading vehicles increased significantly. An estimate of fatal crash reduction was derived by considering type and speed of the road, urbanization, and ambient lighting condition. The estimate is a 4-percent reduction in crashes where ABS could potentially be effective, or about a 2-percent reduction in all fatal crash involvements. The result is not statistically significant.

The primary methodology used in this report is based on comparing a response group of crash involvements where ABS could be influential to a control group of crash involvements where ABS should be irrelevant. In arriving at final estimates of the effectiveness, consideration was given to vehicle age, speed of the road, locality, and ambient lighting. However, differences in driving patterns and vehicle maintenance must be assumed equal, to the extent that the control group and response group crashes are affected equally by uncontrollable factors. There may also be deficiencies in what is provided to NHTSA's State Data System, in terms of all eligible crashes being reported and in terms of the accuracy of the information that is provided. It is not known if reporting varies according to any influential criterion. The influence of other technologies that have appeared in the most recent model years are also not taken into consideration.

Table of Contents

List of Abbreviations

ABS	Antilock brake system
CY	Calendar year
df	Degrees of freedom
DOT	United States Department of Transportation
FARS	Fatality Analysis Reporting System, a census of fatal crashes in the United States since 1975
FMCSA	Federal Motor Carrier Safety Administration
FMVSS	Federal Motor Vehicle Safety Standard
GVWR	Gross vehicle weight rating, specified by the manufacturer, equals the vehicle's curb weight plus maximum recommended loading
LTV	Light Trucks and Vans, includes pickup trucks, SUVs, minivans, and full-size vans
MY	Model year
NHTSA	National Highway Traffic Safety Administration
OR	Odds ratio
SAS	Statistical analysis software produced by SAS Institute, Inc.
SDS	NHTSA's State Data System
SUV	Sport utility vehicle
VIN	Vehicle Identification Number

Introduction

This report analyzes the impact of Federal Motor Vehicle Safety Standard No. 121, which mandates antilock braking systems on virtually all new air-braked vehicles. ABS is required on tractors manufactured on or after March 1, 1997, and air-braked semi-trailers and single-unit trucks manufactured on or after March 1, 1998.

Crash data from NHTSA's State Data System and Fatality Analysis Reporting System are used to determine the effectiveness of antilock brakes in heavy tractor-trailer combination vehicles. The report is organized as follows:

- State data from Florida and a special data collection project in North Carolina are used to determine whether the tractor or the trailer is the more influential portion of a combination vehicle. That is, four possibilities are considered – both tractor and trailer **do not** have ABS; only the trailer has ABS; only the tractor has ABS; both the tractor and trailer **do** have ABS.
- Additional State data sources are then incorporated where information on only the tractor is available. This provides a preliminary estimate of the effectiveness in terms of all levels of crash severity.
- Next, the preliminary effectiveness in fatal crash involvements is determined using FARS, based on the tractor ABS equipment.
- The effectiveness derived from FARS is re-considered based on the type of road and locality where the crash occurred. This step helps account for different exposure patterns for older, non-ABS-equipped vehicles and newer, ABS-equipped vehicles.
- Finally, the results from the State data are revisited to arrive at a final estimate of the reduction in all crash involvements. The age of the tractor is controlled for in this stage of analysis.
- Limitations are presented – namely, tractors prior to the 1997 mandate may have had ABS installed voluntarily and tractors produced after the 1997 mandate may have had ABS not functioning properly.

Research findings

In the Final Economic Assessment for the updated FMVSS No. 121,[1] data from an earlier German study (1984) were used to estimate the effectiveness of ABS. Engineers from NHTSA re-examined the crash reports of 177 crashes involving 182 heavy vehicles. It was determined that combination vehicles would have been involved in 8.86 percent fewer crashes if they had been equipped with ABS. This was not a statistical analysis, and there was no possibility to compare crash involvements to a set of vehicles with ABS at that time.

---

[1] *Final Economic Assessment - Final Rules FMVSS Nos. 105 & 121: Stability and Control During Braking Requirements and Reinstatement of Stopping Distance Requirements for Medium and Heavy Vehicles* (1995); available as document number 22 in docket FHWA-97-2318 from regulations.gov

NHTSA recently published a statistical analysis of ABS effectiveness in light vehicles.[2] The analysis is based on passenger vehicle makes and models that switched from being not equipped with ABS in one model year to being equipped with ABS in the following model year. In some cases, makes and models with optional ABS were included if the percent of vehicles sold with ABS was close to zero or 100 percent.

Involvement in fatal crashes was analyzed using FARS from 1995 to 2007. The net effects were a 1-percent decrease in fatal crash involvement for passenger cars and a 1-percent increase for LTVs. Neither effect was significantly different from zero.

A control group of crashes was constructed based on situations where ABS should not be influential – vehicles moving less than 10 mph, backing up, parking or unparking, struck in the rear, and those not-at-fault (i.e., not "culpable") in other multivehicle crashes. All other crashes are considered part of the response group of crashes that ABS could influence. Single-vehicle response-group crashes include the following: run-off-road with roll-over; run-off-road with side-impact into a fixed object; collisions with pedestrians, bicycles, animals, or other nonmotorists. Multivehicle response vehicles include those which are front-impactors with the rear of another vehicle and those considered culpable.

For the various sub-types of response-group crashes, some statistically significant results were found. Run-off-road crashes were more likely for ABS-equipped passenger cars (9 percent increase), primarily due to a 30 percent increase in side impacts with fixed objects. On the other hand, a decrease of 13 percent is reported for collisions with pedestrians, bicycles, and animals.

The effectiveness of ABS in nonfatal crashes was calculated using General Estimates System of the National Automotive Sampling System. The control and response groups were selected to be identical to the FARS analysis. The overall estimate for passenger cars is a 6 percent reduction in crashes for ABS-equipped passenger cars and 8 percent for ABS-equipped LTVs. Similar to the findings using FARS, the results are composed of several counter-acting effects. Passenger cars and LTVs have similar results when investigated by crash type. Some results are listed below:
- an increase in run-off-road side impacts with fixed objects (20% increase for cars, 9% for LTVs);
- reductions in first-event rollovers (3% reduction for cars, 17% for LTVs) and other run-off-road crashes (5% and 15%);
- an increase in pedestrian/bicycle/animal collisions (8% increase for cars, 42% for LTVs);
- a reduction in multivehicle culpable involvements (17% reduction for cars, 20% for LTVs).

---

[2] Kahane, C. J. (2009). *The Long-Term Effect of ABS in Passenger Cars and LTVs*. DOT HS 811 182. Washington, DC: National Highway Traffic Safety Administration. http://www-nrd.nhtsa.dot.gov/Pubs/811182.PDF

Preliminary analysis of all crashes

The ABS equipment of the tractor is determined from the model year. The model year given in the State databases is checked against the model-year character in the VIN. Generally, the two values agree. If the VIN-derived model year differs from the recorded model year, the VIN-derived model year is used so long as it is valid (i.e., $1980 \leq model\ year \leq$ year of crash +1). If the VIN-derived model year is not valid, the model year in the datasets is used. If neither the model year variable nor the VIN can provide a model year, the vehicle is excluded from analysis. In States where information about the trailer is available, the same rules are followed to determine the trailer model year.

The list below shows the number of tractor-trailer vehicles in the Florida data according to the source of the model year. There is a high degree of agreement between the model year as recorded in the data and that derived from the VIN. For tractors, there is agreement in 87 percent of the vehicles, and there is agreement for 77 percent of the trailers.

	Tractors	Trailers
Model Year in data agrees with VIN Model Year	43,104	34,475
VIN Model Year is used	5,343	7,286
Model Year in data is used	961	3,131

The tractor is assumed to have ABS if the model year is 1998 or newer, according to the FMVSS 121. Any tractor of model year 1996 or older is assumed to not have ABS. The extent of pre-mandate ABS is not clear. The model year 1997 is excluded because the mandate applies to vehicles manufactured on or after March 1, 1997.

The ABS mandate that applied to trailers took effect one year later. All trailers of model year 1999 and newer are assumed to be ABS-equipped and all trailers of model year 1997 and earlier are assumed not to be ABS-equipped. A trailer with a model year of 1998 is considered to have unknown ABS equipment. No manufacturer-specific information has been obtained.

Crash classification using Florida State data as a model

The State data for Florida is described in detail. The analogous steps were carried out with the other States whose results are shown. Due to differences in data reporting and variable definitions, the details from Florida cannot be applied directly to other States.

In Florida, tractor-trailer combinations are identified using two variables. First is the vehicle type (VEH_TYPE). Two values are included, shown below. A vehicle with one of those two vehicle types must also be hauling a trailer of one of the three types listed below. Other possible values include boat or utility trailers, which are not appropriate for analyses of the typical tractor-trailer combination. When a trailer is connected, the difference between type "05" and "06" may be a data-collection inconsistency, perhaps due to the design of the police report form. For calendar years 2003 onwards, approximately one-fourth of the combination vehicles included in this analysis are of the "05" type, and it may bias the results to exclude such a high percentage.

VEH_TYPE	05	Heavy truck (two or more rear axles)
	06	Truck—tractor (cab—bobtail)
TRAILER	01	Single semi-trailer
	03	Tank trailer
	04	Saddle mount/flatbed

The classification of crash involvements consists first in identifying situations where the ABS is potentially effective or influential, i.e., when stopping more quickly or maintaining better control of the vehicle during stopping would conceivably have prevented the crash. The crash involvements where ABS is hypothesized to be influential are compared to a contrasting group of crashes where ABS should not be influential, referred to as the *Control Group* crashes.

The first step in crash classification looks for incidents where vehicles are moving slowly. In Florida, this is based on either the vehicle's speed or the speed limit where the crash occurred; the criteria are shown below. When a vehicle is moving slowly, the ABS is not likely to activate, and even if it activates is unlikely to significantly alter stopping performance. For this analysis, "slow" is considered 10 mph or less, though the precise activation thresholds may vary according to ABS manufacturer.

.

Slow	$1 \leq$	SPEED	≤ 10
	$1 \leq$	SPDLIM	≤ 10

Some other vehicles are excluded based on the maneuver that would indicate slow movement. The final item, for driverless/runaway vehicles, is also specified for the obvious reason that such a vehicle would not be able to use the brakes.

Other exclusions	VEH_MAN1	07	entering or leaving parking
		08	properly parked
		09	improperly parked
		12	driverless/runaway

When vehicles are slow-moving or meet one of the other criteria, they are eligible for inclusion in the control group but cannot enter the group of crashes that ABS could prevent. For example, a vehicle could be slow-moving while turning and be rear-ended by another vehicle traveling at the prevailing road speed. The slow-moving vehicle would be in the control group and the vehicle that hit the turning vehicle would be in the response group. If all vehicles in a crash are slow-moving or meet one of the other exclusions above, the crash is excluded, even as a source of control-group vehicles – these could be crashes that occur in a parking lot where both vehicles are moving slowly or where one backs into another.

The first group of crashes to be identified that ABS could conceivably prevent are single-vehicle crashes. Several different scenarios are classified, as follows:

- Run-off-road can occur when a vehicle loses control during an avoidance maneuver. These crashes are identified by the road-relation data element (RD_REL = 2). It is further noted whether the vehicle overturned (IMPACT = 19).
- Collisions with animals, pedestrians, and bicycles may be avoidable if ABS shortens stopping distance or maintains steerability. The vehicle event variables are used to identify these collisions (VEVENT1 or VEVENT2 or VEVENT3 or VEVENT4 = 10 or 11 or 12 or 15).
- Jackknife crashes result from wheel lockup of the tractor or trailer, causing the tractor and trailer to skid into a folded position (VEVENT1 or VEVENT2 or VEVENT3 or VEVENT4 = 33).
- Any single vehicle crashes that do not meet the above criteria are classified as Other Single Vehicle Crashes. Most of these are fixed-object impacts or overturns that are not classified as run-off-road according to the first rule above.

Crashes with more than one vehicle are classified next. The steps are the following:
- Vehicles that are impacted in the rear are in the control group (IMPACT = 6 or 7 or 8 or 9 or 10). In States which provide more than one impact zone variable, only the initial point of impact is considered. A vehicle hit in the rear would not presumably have benefited from shorter stopping distances or better handling provided by ABS.
- When there are exactly two vehicles in the crash, the vehicle which impacts with its front (IMPACT = 1 or 2 or 3 or 13 or 14) into the rear of the other is part of the response group.
- If the point of impact does not identify a vehicle as part of the control (rear-impacted) or response (rear-impacting) groups, the vehicle at-fault classification is used. A vehicle considered to be at-fault (VEH_FLT = 1) is part of the response group, and a vehicle not-at-fault is part of the control group. This classification cannot be perfect – for example, ABS could benefit a tractor-trailer to maintain control in a crash where it was cut-off by a car, through no fault of the driver of the tractor-trailer. Fault status is used because it is widely available in the NHTSA's State Data System, thus allowing consistency throughout the analyses.

Consideration of both tractor and trailer ABS

Data from seven States is included in the analysis. The data must have vehicle-type information to identify tractor-trailers and either the VIN or model year to classify the ABS equipment. The variables describing the crash circumstances must be approximately similar to those described in detail above for Florida – i.e., speed or vehicle maneuvers to identify slow-moving vehicles; vehicle events to classify single-vehicle crashes; impact zone and vehicle fault for multivehicle crashes.

Information about the trailer is not available in very many states – most list only whether a trailer is attached to a vehicle. Some States allow multiple trailers, but these are

5

excluded when separate vehicle type categories make a distinction between single and multiple trailers (e.g., Ohio and Wisconsin). Only Florida has either the model year or VIN for the trailer. Additionally, a special data collection project in North Carolina was conducted by NHTSA in cooperation with the State Highway Patrol, from which some data can be used. Below are listed the States used in the analysis.

Tractor and Trailer model year available	Calendar years of data
Florida	1998 – 2007
North Carolina	2005 – 2007 (special NHTSA study)
Tractor-only model year available	
Missouri	2002 – 2006
Ohio	2000 – 2006
Wisconsin	1998 – 2006
Georgia	1998 – 2006
North Carolina	2003 – 2006 (State data)

Table 1 contains the absolute number of crashes in Florida for calendar years 1998 to 2007. The four columns show the possible combinations of Tractor and Trailer ABS equipment. At left, listed as "NO" and "NO" are combinations vehicles where the tractor has a model year of 1996 or older and the trailer has a model year of 1997 or older.

Table 1: Number of crashes involvements for tractors and trailers based on ABS equipment, Florida State data

	TRACTOR ABS	NO	NO	YES	YES
	TRAILER ABS	NO	YES	NO	YES
Control group		5354	1879	4867	5837
All response group		3201	1138	2490	3011
All single vehicle		1201	399	759	809
Run-off-road					
Overturn		158	40	132	121
Other		373	140	233	273
Ped/Bike/Animal		43	23	57	54
Jackknife		84	27	18	32
Other		543	169	319	329
All multi-vehicle response crashes		2000	739	1731	2202
Impacting in two-vehicle front-to-rear		811	329	823	1069
At-fault in other multi-vehicle collisions		1189	410	908	1133

Table 2 shows the percent reduction in each of the crash types relative to the combination NO-NO vehicle. At right, in smaller font, are the values for the χ^2 test of significance. When the χ^2 is greater than 3.84, the percentage reported at left is significantly different from zero and printed in **bold black** when positive and **bold red** when negative.

Table 2: Reduction in response group crashes based on tractor and trailer ABS equipment, Florida state data

	TRACTOR ABS NO / TRAILER ABS YES	TRACTOR ABS YES / TRAILER ABS NO	TRACTOR ABS YES / TRAILER ABS YES	χ2		
All response group	-1%	**14%**	**14%**	0.1	22.0	21.7
All single vehicle	5%	30%	38%	0.7	52.3	96.7
Run-off-road						
Overturn	28%	8%	30%	3.4	0.5	8.4
Other	-7%	**31%**	33%	0.4	19.3	24.0
Ped/Bike/Animal	-52%	-46%	-15%	2.7	3.5	0.5
Jackknife	8%	**76%**	**65%**	0.2	36.4	27.8
Other	11%	**35%**	**44%**	1.7	35.9	67.4
All multi-vehicle response crashes	-5%	5%	-1%	1.0	1.6	0.1
Impacting in two-vehicle front-to-rear	-16%	-12%	-21%	4.2	4.3	14.3
At-fault in other multi-vehicle collisions	2%	**16%**	**13%**	0.1	13.0	8.7
Overall reduction in crashes	0%	5%	5%			

The percent reduction is calculated by subtracting the Odds Ratio from one. The basic method is to construct a 2×2 table for each row, relative to the Control Group. The case of All Response Group is illustrated below, for the YES-YES vehicles.

	NO-NO	YES-YES
Control Group	5354	5837
All Response Group	3202	3011

A general table can be defined symbolically as follows, using n to represent the counts and the subscripts 1 and 2 to identify the appropriate row and column, in that order.

	NO-NO	YES-YES
Control Group	n_{11}	n_{12}
All Response Group	n_{21}	n_{22}

The relevant computations to arrive at the effectiveness are illustrated below, both symbolically and numerically.

Odds for NO-NO column $n_{21} \div n_{11} = 3202 \div 5354 = 0.598$
Odds for YES-YES column $n_{22} \div n_{12} = 3011 \div 5837 = 0.516$

Odds Ratio for YES-YES relative to NO-NO
$$[n_{22} \div n_{12}] \div [n_{21} \div n_{11}] =$$
$$[3011 \div 5837] \div [3202 \div 5354] = 0.516 \div 0.598 = 0.863$$

Reduction in Odds Ratio
$$1 - 0.863 = 0.137 = 13.7\% = \text{Effectiveness of ABS}$$

Overall reduction in crashes
$$13.7\% \times [3011 \div (3011 + 5837)] = 5\%$$

The statistical significance test on the odds ratio is a χ^2 with one degree of freedom. The formula is shown below continuing with the example above.

$$\chi^2 = \quad [\text{ difference in the product of diagonal elements }]^2 \times [\text{sum of all elements}]$$
$$\div [\text{product of all four non-diagonal sums}]$$

$$\chi^2 = \quad [n_{11} \times n_{22} - n_{12} \times n_{21}]^2 \times [n_{11} + n_{22} + n_{12} + n_{21}] \div$$
$$[(n_{11} + n_{12}) \times (n_{12} + n_{22}) \times (n_{22} + n_{21}) \times (n_{21} + n_{11})]$$

$$\chi^2 = \quad [5354 \times 3011 - 5837 \times 3202]^2 \times [5354 + 3011 + 5837 + 3202] \div$$
$$[(5354 + 3202) \times (5354 + 5837) \times (5837 + 3011) \times (3011 + 3202)]$$

$$\text{degrees of freedom} \quad = (\text{number of rows} - 1) \times (\text{number of columns} - 1)$$
$$= (2 - 1) \times (2 - 1) = 1 \times 1 = 1$$

The reduction in the all response-group crashes relative to the control-group crashes is listed at the top of the table. The last line represents the reduction in the total number of response *and* control group crashes. This value must be interpreted in the context of NHTSA's State Data System because the relative column percentages vary by State, according to the crash reporting criteria and the available crash classification variables.

Of special interest for ABS are crashes that occur on wet road surfaces because ABS is intended to mitigate wheel-lock that would lead to slipping or skidding. In Florida, crashes on wet roads are identified at the crash-level according to the following rule:

RD_SUR	02	Wet
	03	Slippery
	04	Icy

Table 3 shows the number of response-group crashes on wet roads. The control group, as shown in Table 1, is not changed because these are crashes where ABS is assumed to be inactive or irrelevant.

Table 3: Number of crashes involvements on wet roads for tractors and trailers based on ABS equipment, Florida State data

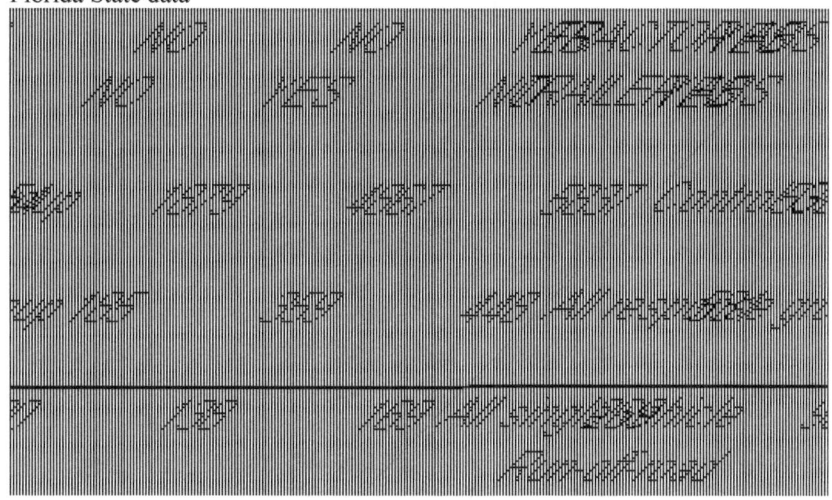

Table 4 shows the crash reduction and associated χ test statistics for the data in Table 3. For both of the Tractor YES columns, the ABS is more effective on wet roads. This is not explicitly true in terms of every crash scenario, but most importantly the reduction is greater for the row All Response Group, as well as for the aggregated All Single Vehicle and All Multivehicle crash scenarios.

Table 4: Reduction in response group crashes on wet roads based on tractor and trailer ABS equipment, Florida state data

TRACTOR ABS TRAILER ABS	NO YES	YES NO	YES YES	χ2		
All response group	1%	26%	23%	0.0	17.4	14.8
All single vehicle Run-off-road	5%	44%	42%	0.1	29.5	31.1
Overturn	29%	1%	-24%	0.5	0.0	0.5
Other	-1%	45%	43%	0.0	14.9	14.9
Ped/Bike/Animal	-42%	-93%	31%	0.2	1.1	0.2
Jackknife	3%	77%	67%	0.0	23.6	19.0
Other	12%	35%	42%	0.3	4.9	8.2
All multi-vehicle response crashes	-3%	7%	2%	0.1	0.6	0.1
Impacting in two-vehicle front-to-rear	-41%	-30%	-29%	3.6	3.4	3.5
At-fault in other multi-vehicle collisions	17%	28%	20%	1.4	7.0	3.7
Overall reduction in crashes	0%	2%	2%			

Table 5 lists the number of crashes for the Control and All Response groups for Florida and North Carolina in terms of the combined tractor and trailer ABS equipment. Florida has a greater amount of data available across the four combinations. The special study in North Carolina did not amass a great number in any of the columns.

Table 5: Number of crashes involvements for tractors and trailers based on ABS equipment, Florida State data and North Carolina special study

		Tractor ABS NO	NO	YES	YES
		Trailer ABS NO	YES	NO	YES
FL	Control group	5354	1879	4867	5837
	All response group	3201	1138	2490	3011
	Wet response group	532	185	360	448
NC	Control group	219	101	399	595
	All response group	346	145	496	831
	Wet response group	62	23	100	138

Table 6 shows the reduction in all response group crashes relative to the control group crashes for the two States where the trailer ABS equipment can be inferred. In Florida, the results show little or no evidence that an ABS-equipped trailer, being hauled by pre-standard tractor, has fewer crash involvements than when neither the tractor nor the trailer is ABS-equipped. Likewise, in Florida, the results show little or no evidence that ABS on the trailer enhances the effect of ABS on the tractor. In North Carolina, the result is not clear – the combination "Tractor YES / Trailer NO" has the highest effectiveness and is the only one of the three that is significantly different from zero, while the two combinations with "Trailer YES" are very similar (9% versus 12%). The results indicate that further analysis can proceed with consideration of only the tractor ABS.

Table 6: Reduction in response group crashes based on tractor and trailer ABS equipment, Florida state data and North Carolina special study

		Tractor ABS NO	YES	YES
		Trailer ABS YES	NO	YES
FL	All response group	-1%	**14%**	**14%**
	Wet response group	1%	**26%**	**23%**
NC	All response group	9%	**21%**	12%
	Wet response group	20%	11%	18%

Consideration of only tractor ABS

Table 7 lists the number of Control Group and All Response crashes for each of the seven States, based only on Tractor ABS. The tallies for Florida are larger than shown above because cases are included which are from the transitional model year 1998 or else have unknown trailer information. The tallies for North Carolina are from the State Data System for calendar years 2003-2006, whereas the above tractor-plus-trailer information was based only on a small fraction of cases from the Special Study.

Table 7: Number of crashes involvements for tractors-trailers based on ABS equipment of the tractor, seven States

		Tractor NO	Tractor YES
FL	Control group	7663	11620
	All response	4599	5937
GA	Control group	20090	25520
	All response	20158	22686
IN	Control group	2485	6946
	All response	4205	10619
MO	Control group	2826	7663
	All response	4956	11726
NC	Control group	2469	6168
	All response	3666	7591
OH	Control group	11187	20340
	All response	14478	22943
WI	Control group	7179	9312
	All response	8559	10010

Confidence intervals are calculated from the standard error of the *log (odds ratio)* according to the following:

$$SE \text{ (log OR)} = \text{sqrt}(\ 1/n_{11} + 1/n_{12} + 1/n_{21} + 1/n_{22}\ )$$

For Florida, the corresponding values are:

$$SE \text{ (log OR)} = \text{sqrt}(\ ^1/_{7663} + ^1/_{4599} + ^1/_{11620} + ^1/_{5937}\ ) = 0.025$$

A symmetric confidence interval is constructed on the *log OR* by adding and subtracting a multiple (Z) of the standard error:

$$\text{Confidence Bounds} = \log OR \pm Z \times SE$$

For a 95 percent confidence interval, the value for Z is 1.96. Continuing with Florida as the example:

$$OR = (5937 \div 11620) \div (4599 \div 7663) = 0.851$$

and

$$\log (OR) = \log(0.851) = -0.161$$

The confidence bounds are thus:

$$-0.161 \pm 1.96 \times 0.025 = (-0.209, -0.113)$$

Raising to the exponent e, these values are re-converted to odds ratios:
exp(-0.209) = 0.81
exp(-0.113) = 0.89

Finally, the confidence bounds on the odds ratios are subtracted from 1 to yield the confidence bounds on the percent reduction:
1- 0.81 = 19%
1- 0.89 = 11%

Table 8 summarizes the percent reduction and statistical significance test based on ABS of the tractor. The median of the seven values is 13 percent. All values are significantly different from zero (all $\chi^2 > 3.84$), and the p-values show that these results are highly significant. The confidence intervals range from an at-worst low-end estimate of 4 percent (Indiana) to an at-best high-end estimate of 22 percent (North Carolina).

Table 8: Reduction in response group crashes and statistical significance test, data from seven States

	Reduction	χ^2 test	Lower CI	Upper CI
FL	15%	43.04 $p < 0.01$	11%	19%
GA	11%	80.31 $p < 0.01$	9%	14%
IN	10%	11.74 $p < 0.01$	4%	15%
MO	13%	24.11 $p < 0.01$	8%	17%
NC	17%	36.30 $p < 0.01$	12%	22%
OH	13%	75.25 $p < 0.01$	10%	16%
WI	10%	23.14 $p < 0.01$	6%	14%

Table 9 shows the results for only response group crashes which occurred on wet roads. The percent reduction and statistical test information are included to the right. Except for Missouri, all results are statistically significant. The median percent reduction is 17 percent. The lowest confidence interval extends to 3 percent (among those with $\chi^2 > 3.84$), while the highest is 32 percent.

Table 9: Number of crash involvements on wet roads with percent reduction and statistical test, data from seven States

		Tractor NO	Tractor YES	Reduction	Chi-Sq	Lower CI	Upper CI
FL	Control group	7663	11620	25%	31.42	17%	32%
	Wet response	759	861		p < 0.01		
GA	Control group	20090	25520	19%	70.01	15%	23%
	Wet response	3468	3555		p < 0.01		
IN	Control group	2485	6946	11%	6.52	3%	18%
	Wet response	939	2339		p = 0.01		
MO	Control group	2826	7663	6%	1.58	-3%	14%
	Wet response	882	2259		p = 0.21		
NC	Control group	2469	6168	18%	13.98	9%	26%
	Wet response	692	1423		p < 0.01		
OH	Control group	11187	20340	8%	10.57	3%	12%
	Wet response	3425	5749		p < 0.01		
WI	Control group	7179	9312	17%	31.27	11%	22%
	Wet response	2186	2350		p < 0.01		

In five of seven states, the effectiveness on wet roads exceeds the effectiveness on all roads. The effectiveness is lower on wet roads for Missouri and Ohio. For comparison, the results are summarized in Table 10. At right, the percentage-point improvement on wet roads ranges from -7 to 10 percent.

Table 10 : Comparison of reduction in control group crashes on all roads versus wet roads

	ALL	WET	WET minus ALL
FL	15%	25%	+10%
GA	11%	19%	+8%
IN	10%	11%	+1%
MO	13%	6%	-7%
NC	17%	18%	+1%
OH	13%	8%	-5%
WI	10%	17%	+7%

Table 11 shows the estimates of effectiveness for each of the classified crash types. Because there are so many comparisons, some results could be statistically significant merely by chance. The results are presented in this way to help in identifying patterns and consistencies in the results.

Table 11 : Reduction in response group crashes for various crash mechanisms, summary of seven States

	FL	GA	IN	MO	NC	OH	WI
All response group	15%	11%	10%	13%	17%	13%	10%
All single vehicle	34%	23%	9%	10%	23%	8%	12%
Run-off-road							
Overturn	14%	26%	33%	21%	53%	16%	6%
Other	33%	26%	2%	6%	18%	11%	15%
Animal/Ped/Bike	-15%	-6%	-24%	-55%	-63%	-79%	-18%
Jackknife	69%	72%	41%	46%	72%	49%	53%
Other	39%	16%	12%	21%	12%	25%	12%
All multi-vehicle response crashes	4%	7%	10%	15%	14%	17%	8%
Impacting in two-vehicle front-to-rear	-10%	5%	9%	8%	9%	5%	13%
At-fault in other multi-vehicle collisions	14%	10%	10%	18%	15%	22%	7%
Overall reduction in crashes	5%	5%	6%	8%	9%	7%	5%

In nearly all cases, the results are in the same direction and comparable in magnitude. The *Animal/Ped/Bike* single-vehicle crash type has a negative effectiveness in all states – that is, tractors with ABS are *more likely* to be involved in crashes with animals, pedestrians, and bicycles relative to the control group crashes. This result is consistent with what will be reported in analysis of FARS data in the next section. The difference, on the other hand, is that these crashes from the State data are predominantly animal impacts (e.g., deer), whereas fatal accidents are primarily pedestrian or bicycle impacts. The reduction in jackknife crashes is the largest effect in each. This result is important because jackknife is one crash mechanism that tractor-trailer ABS is intended to mitigate, as cited in the Final Economic Assessment[1].

The numbers of crashes for all crash mechanisms are shown in Table 12. This aids in interpretation of Table 11. For example, although there is a relative increase in the Animal/Ped/Bike category, these types of crashes are not common.

Table 12: Absolute number of crashes for various crash mechanisms, based on Tractor ABS, data from seven States

	FL		GA		IN		MO	
	ABS NO	ABS YES	ABS NO	ABS YES	ABS NO	ABS YES	ABS NO	ABS YES
Control	7663	11620	20090	25520	2485	6946	2826	7663
All response group	4599	5937	20158	22686	4205	10619	4956	11726
All single vehicle	1700	1708	5188	5082	1490	3797	1858	4550
Run-off-road								
Overturn	210	274	819	774	53	99	193	412
Other	550	557	2084	1966	213	581	1012	2581
Animal/Ped/Bike	68	119	306	412	181	627	90	379
Jackknife	118	56	246	86	81	134	42	61
Other	754	702	1733	1844	962	2356	521	1117
All multi-vehicle response crashes	2899	4229	14970	17604	2715	6822	3098	7176
Impacting in two-vehicle front-to-rear	1207	2016	8070	9747	702	1779	960	2399
At-fault in other multi-vehicle collisions	1692	2213	6900	7857	2013	5043	2138	4777

Table 12a: Absolute number of crashes for various crash mechanisms, based on Tractor ABS, data from seven States (continued)

	NC		OH		WI	
	ABS NO	ABS YES	ABS NO	ABS YES	ABS NO	ABS YES
Control	2469	6168	11187	20340	7179	9312
All response group	3666	7591	14478	22943	8559	10010
All single vehicle	1310	2504	6469	10823	4223	4828
Run-off-road						
Overturn	285	336	446	683	262	321
Other	388	798	2271	3694	1255	1385
Animal/Ped/Bike	48	195	760	2474	312	477
Jackknife	83	58	285	263	161	98
Other	506	1117	2707	3709	2233	2547
All multi-vehicle response crashes	2356	5087	8009	12120	4336	5182
Impacting in two-vehicle front-to-rear	664	1512	2576	4461	893	1009
At-fault in other multi-vehicle collisions	1692	3575	5433	7659	3443	4173

The results of Table 11 are summarized in Table 13 in terms of the median (MED), minimum (MIN), and maximum (MAX) effectiveness calculations. The combined results in terms of single-vehicle and multiple-vehicle crashes are very close (median reductions 12% versus 10%) and correspondingly very close to the overall result for reduction in all response group crashes (13% reduction).

Table 13: Summary of reductions for crash mechanisms in Table 11

	MED	MIN	MAX
All response group	13%	10%	17%
All single vehicle	12%	8%	34%
Run-off-road			
Overturn	21%	6%	53%
Other	15%	2%	33%
Animal/Ped/Bike	-24%	-79%	-6%
Jackknife	53%	41%	72%
Other	16%	12%	39%
All multi-vehicle response crashes	10%	4%	17%
Impacting in two-vehicle front-to-rear	8%	-10%	13%
At-fault in other multi-vehicle collisions	14%	7%	22%
Overall reduction in crashes	7%	5%	9%

Finally, the results of the χ^2 tests are shown in Table 14. Most values are beyond the critical value of 3.84, which represents a significant effect at $p < 0.05$. Values shown in **bold black** are significant reductions, and values shown in **bold red** are significant increases. There is no significance test on the lower row "Overall reduction in crashes" from Table 11 because this is a derived from both response and control groups, rather than a 2×2 table calculation.

Table 14: χ^2 statistics for crash reductions in Table 11

	FL	GA	IN	MO	NC	OH	WI
All response group	43.04	80.31	10.91	24.11	36.30	75.25	23.14
All single vehicle	122.94	141.59	5.77	8.36	41.54	17.95	23.09
Run-off-road							
Overturn	2.62	33.84	5.58	7.10	83.04	7.68	0.45
Other	43.04	82.63	0.09	2.02	8.62	14.60	14.81
Animal/Ped/Bike	0.88	0.59	6.02	13.82	9.08	186.74	4.88
Jackknife	56.64	120.63	13.81	9.89	61.68	64.01	35.95
Other	81.47	26.03	8.70	16.78	4.45	103.34	15.22
All multi-vehicle response crashes	1.88	28.01	10.70	26.25	17.91	97.48	10.01
Impacting in two-vehicle front-to-rear	6.06	8.06	3.79	3.43	3.14	3.16	8.09
At-fault in other multi-vehicle collisions	17.38	33.13	9.67	32.31	19.53	143.18	5.93

Preliminary analysis of fatal crashes

Description of vehicle and crash identification

Crashes which result in a fatality were analyzed using FARS, for the calendar years 1998 to 2008. Tractor-trailers are identified in FARS using the BODY_TYP variable. These vehicles are further restricted to those hauling one trailer using the TOW_VEH variable.

Calendar Year of Data	Variables	Description
1991 and later	BODY_TYP = 66	Truck/Tractor
2004 and later	TOW_VEH = 1	Yes, One Trailer
1983 to 2003	TOW_VEH = 1	Yes, One Trailing Unit

Crash classification resembles the procedure described above for the Florida data. Slow-moving vehicles are identified as those with travel speeds below 10 miles per hour or as those involved in crashes where the speed limit is less than 10 miles per hour.

Slow
Vehicle level	$1 \leq$	TRAV_SP	≤ 10	(Travel speed)
Accident level	$1 \leq$	SP_LIMIT	≤ 10	(Speed limit)

Otherwise, vehicles can be identified as likely to be slow-moving using the vehicle maneuver (VEH_MAN) variable. Shown below are several further exclusions which prevent vehicles from entering the response group.

VEH_MAN	4	Stopped in Traffic Lane
	6	Leaving a Parked Position
	7	Parked
	8	Entering a Parked Position

Single-vehicle crashes are identified first based on whether the vehicle ran off the road, using the variable REL_ROAD as indicated below. Vehicles which over-turned or rolled over are classified using the harmful event variable (HARM_EV = 1)

Run-off-road	REL_ROAD	2	Shoulder
		3	Median
		4	Roadside
		5	Outside Trafficway / Outside Right-of-way
		6	Off Roadway – Location Unknown
		7	Parking Lane/Zone
		8	Gore
		10	Separator
Overturn	HARM_EV	1	First-event Overturn/Rollover
Other			other cases where REL_ROAD equals 2,3,4,5,6,7,8 or 10

The harmful event is also used to identify pedestrian, bicycle, and animal collisions, as indicated below. Ninety-nine percent of these crashes are classified as on-road.

Animal/ped/bike	HARM_EV	
	8	Pedestrian
	9	Pedalcycle
	11	Animal
	15	Nonmotorist on Personal Conveyance
	49	Ridden Animal or Animal-Drawn Conveyance

Jackknife crashes are identified in two stages. First, prior to classifying off-road crashes, the jackknife variable (J_KNIFE) is inspected for crashes where a jackknife is recorded as the first event. This ensures that a vehicle that jackknifed on the road but later ran off would be classified as jackknife as opposed to `other run off road`. The second criterion for jackknife crashes is based simply on the harmful event and is available only from 2004 onward.

Jackknife		J_KNIFE	2	First Event
	or	HARM_EV	51	Jackknife

Overturn crashes are also included that did not involve run-off-road. The same criterion is used as for off-road overturns (HARM_EV = 1).

The first type of multivehicle crashes identified is rear-end crashes – one vehicle impacts with its front portion into the rear portion of another vehicle. The manner-of-collision variable is used to help ensure that no peculiar crash scenarios are included. The vehicle hit in the rear is part of the control group (IMPACT1 = 5 or 6 or 7), and the vehicle which impacts with its front is in the response group (IMPACT1 = 1 or 11 or 12; and manner-of-collision MAN_COLL = 1).

Other involvements in multivehicle crashes are classified based on by assigning fault according to either of two variables. FARS identifies up to three violation charges (VIOLCHG1, VIOLCHG2, VIOLCHG3). The broad categories classified as at-fault are shown below.

VIOLCHG1 or VIOLCHG2 or VIOLCHG3
1 to 9   RECKLESS/CARELESS/HIT-AND-RUN OFFENSES
11 to 19 IMPAIRMENT OFFENSES
21 to 29 SPEED-RELATED OFFENSES
31 to 39 RULES OF THE ROAD – TRAFFIC SIGN & SIGNALS
41 to 49 RULES OF THE ROAD – TURNING, YIELDING, SIGNALING
51 to 59 RULES OF THE ROAD – WRONG SIDE, PASSING & FOLLOWING
61 to 69 RULES OF THE ROAD – LANE USAGE

The driver contributing circumstances are also used to identify at-fault vehicles. These values are approximately comparable to those held in the violation charges above, and they are included to help ensure full identification of at-fault vehicles.

The excluded categories are *NON-MOVING – LICENSE & REGISTRATION VIOLATIONS* and *EQUIPMENT* and *OTHER VIOLATIONS*.

DR_CF1 or DR_CF2 or DR_CF3 or DR_CF4

3	Emotional (e.g., Depression, Angry, Disturbed)
6	Inattentive/Careless
8	Road Rage/Aggressive Driving (since 2004)
26	Following Improperly
27	Improper or Erratic Lane Changing
28	Failure to Keep in Proper Lane or Running off Road (1982-1999)
28	Failure to Keep in Proper Lane (since 2000)
30	Making Improper Entry to or Exit from Trafficway
31	Starting or Backing Improperly
33	Passing Where Prohibited by Posted Signs, Pavement Markings, Hill or Curve, or School Bus Displaying Warning Not to Pass
35	Passing with Insufficient Distance or Inadequate Visibility or Failing to Yield to Overtaking Vehicle
36	Operating the Vehicle in an Erratic, Reckless, Careless or Negligent Manner or Operating at Erratic or Suddenly Changing Speeds
38	Failure to Yield Right of Way
39	Failure to Obey Traffic Actual Signs, Traffic Control Devices or Traffic Officers, Failure to Observe Safety Zone Traffic Laws
44	Driving too Fast for Conditions or in Excess of Posted Speed Limit
46	Operating at Erratic or Suddenly Changing Speeds (1982 - 1994)
46	Racing (since 1998)
47	Making Right Turn from Left-Turn Lane or Making Left Turn from Right-Turn Lane
48	Making Improper Turn
50	Driving Wrong Way on One-Way Trafficway
51	Driving on Wrong Side of Road (Intentionally or Unintentionally)
57	Locked Wheel
58	Over Correcting
79	AVOIDING, SWERVING, OR SLIDING DUE TO Slippery or Loose Surface
87	AVOIDING, SWERVING, OR SLIDING DUE TO Ice, Water, Snow, Slush, Sand, Dirt, Oil, Wet Leaves on Road

Initial estimates of ABS effectiveness

Table 15 shows the number of control group and response group crashes based on the ABS equipment of the tractor. The overall result is a not statistically significant difference from zero and is in fact unexpectedly in the negative direction, i.e., tractors equipped with ABS were involved in crashes slightly more often than non-ABS-equipped tractors, relative to the control group crashes. At the next level of detail, single vehicle crashes show a slight negative effect (4% increase), though the result is not statistically significant. All multivehicle response crashes show a slight increase as well (1% increase), though again the result is not statistically significant.

19

Table 15: Reduction in response group crashes for various crash mechanisms, FARS data

	ABS NO	ABS YES	Percent Reduction	χ^2
Control group	8162	12014		
All response group	4026	6073	-2%	0.97
All single vehicle	1729	2647	-4%	1.33
Run-off-road				
Overturn	257	289	24%	9.65
Other	665	1146	-17%	9.64
Ped/Bike/Animal	538	868	-10%	2.62
Jackknife	78	96	16%	1.37
On-the-road Overturn	121	163	8%	0.54
Other	70	85	18%	1.41
All multi-vehicle response crashes	2297	3426	-1%	0.19
Impacting in two-vehicle front-to-rear	401	849	-44%	34.40
At-fault in other multi-vehicle collisions	1896	2577	8%	5.67
Overall reduction in crashes			-1%	

Delving further, single vehicle run-off-road crashes contain opposing effects – first-event overturn crashes are reduced at a statistically significant level (24% reduction), but the result is countered by a statistically significant increase in run-off-road crashes where the vehicle does not overturn (e.g., impacting a tree or other fixed object). A negative effect in single-vehicle crashes is also observed in collisions with pedestrians, bicycles, and animals (10% increase, not statistically significant). First-event jackknife crashes are reduced (16% reduction), although the result is not statistically significant owing to a small number of crashes. Overturn crashes that are not classified as run-off-road show a slight reduction (8%, not statistically significant), as does the small catch-all category `other` (18% reduction, not statistically significant).

Multiple vehicle response-group crashes similarly fall into two categories with opposite results. There is a statistically significant 44 percent increase with ABS of involvements as the striking vehicles in two-vehicle front-to-rear collisions. In multivehicle crashes that are not two-vehicle front-to-rear collisions, tractors with ABS are less likely to be at-fault (statistically significant 8% reduction).

The preceding analysis classified the vehicles in front-to-rear collisions according to their damage location alone, not their "at-fault" status. Only 12 percent of vehicles hit in the rear are considered at-fault, whereas 64 percent of vehicles which hit another vehicle in the rear are at-fault. That is directionally consistent with intuition, but falls short of the close-to-zero and close-to-100 percent that might have been expected – FARS often does not necessarily assign fault to a vehicle in each crash; even in single vehicle crashes, the tractor-trailer is only considered at-fault in 51 percent of the crashes. This suggests that classification based on the impact zone is more thorough for front-to-rear crashes than a classification based on fault.

A portion of the previous table (Table 15) is reproduced in Table 16 where all multivehicle crashes have been re-classified strictly on at-fault status. Tractors with ABS show a large increase in involvements as the impacting vehicle while at-fault (42% increase). On the other hand, ABS is considered effective at reducing the likelihood of being at-fault when rear impacted (23% reduction). The mechanism for this is unclear. The net result is a 2 percent reduction in all response group crashes, compared to a 2 percent increase in Table 15 – neither method arrives at a result that is significantly different from zero.

Table 16: Consideration of all multivehicle crashes strictly on at-fault status

At-fault when rear impacted	279	321	23% reduction	$\chi^2 = 9.95$
Impacting and at-fault	255	543	42% increase	$\chi^2 = 21.20$
Other at-fault **	1,896	2,577	9% reduction	$\chi^2 = 8.02$
All multivehicle	2,430	3,441	5% reduction	$\chi^2 = 3.19$
All single vehicle **	1,729	2,647	2% increase	$\chi^2 = 0.50$
All response	4,159	6,088	2% reduction	$\chi^2 = 0.70$
Control Group	8,029	11,999		

** note: the counts on these rows do not change, but they are included because the slight change in the Control Group counts produces a slight change in the percent reduction and the χ^2.

The response group crashes are further classified as whether they occurred on wet roads. The variable surface condition (SUR_COND) is used, according to the following:

SUR_COND	2	Wet
	3	Snow or Slush
	4	Ice/Frost (2007 and later)
	4	Ice (1975 to 2006)
	5	Sand, Dirt, Mud, Gravel (2007 and later)
	5	Sand, Dirt, Oil (1975 to 2006)

(note: category 5 is < 1% of others)

Crashes on wet roads comprise 16 percent of the response group for non-ABS-equipped tractors and 18 percent for ABS-equipped tractors. There is a 16 percent increase in all response group crashes relative to the same control group as presented in Table 15. This differs from the State data analyses which showed that ABS is more effective on wet roads in all levels of crash severity (Table 10).

Final estimation stages

Fatal crashes by locality and type of road

The results thus far are not encouraging in terms of the effectiveness of ABS at reducing fatal crash involvements. In crashes that are severe enough to result in a fatality, there are quite possibly multiple factors. The benefits of ABS (e.g., shorter stopping distance or better control) may merely change the mechanism of the crash.

Table 17 shows the number of crashes according to two characteristics of the location. First is whether the locality was rural. The FARS variable ROAD_FNC is used to make this determination – the values 1, 2, 3, 4, 5, 6, and 9 are *rural*; the values 11, 12, 13, 14, 15, 16, and 19 are *non-rural*. Second, the roads are classified as *high-speed* or not according to the ROAD_FNC interstate (values 1 or 11) *or* a speed limit of 55 or higher.

The response-group crashes are shown for single-vehicle and multivehicle crashes. The first set labeled `All localities & road speeds` is a repeat of Table 15, for comparison. Within each road and locality, the calculations for percent reduction and the associated χ^2 are based on the control group for that combination.

For both not high-speed localities (rural and non-rural), there is a statistically significant reduction in crash involvement for tractors equipped with ABS. The subsets for single-vehicle and multivehicle crashes are similar to each other and to the overall reduction. On roads that are not high speed, there is greater potential for avoiding a crash (as opposed to crashing in some other manner) – the travel speed would be lower and less surrounding traffic would be present on secondary streets in rural areas and side streets in urban areas.

The reduction in crashes on low-speed roads is countered by a statistically significant increase in crashes on rural high-speed roads. Although the percentage increases are lower, these roads account for the largest proportion of the crashes and thus make the overall results slightly negative (though insignificant).

Table 17: Number of crashes and reduction for ABS-equipped tractors according to type of locality and speed of road, FARS data

	NO ABS	YES ABS	% Reduction	χ2
All localities & road speeds				
Control group	8162	12014		
All response group	4026	6073	-2%	0.97
Single vehicle	1608	2484	-5%	1.90
Multi-vehicle	2297	3426	-1%	0.19
Rural & not high-speed road				
Control group	869	959		
All response group	443	373	24%	10.29
Single vehicle	177	168	14%	1.65
Multi-vehicle	235	183	29%	10.26
Rural & high-speed road				
Control group	5023	7286		
All response group	2341	3778	-11%	11.07
Single vehicle	870	1471	-17%	10.86
Multi-vehicle	1406	2200	-8%	3.82
Non-rural & high-speed road				
Control group	1359	2472		
All response group	735	1357	-1%	0.07
Single vehicle	300	556	-2%	0.06
Multi-vehicle	420	774	-1%	0.04
Non-rural & not high-speed road				
Control group	868	1231		
All response group	489	537	23%	11.16
Single vehicle	253	276	23%	7.24
Multi-vehicle	228	256	21%	5.33

The results for the two sets above that are not on high-speed roads are very similar. In Table 18 , these two categories have been combined for the result on all roads that are not high-speed.

Table 18: Number of crashes and reduction for ABS-equipped tractors on roads that are not high-speed, FARS data

	NO ABS	YES ABS	% Reduction	χ2
All not high-speed				
Control group	1737	2190		
All response group	932	910	23%	20.43
Single vehicle	430	444	18%	7.12
Multi-vehicle	463	439	25%	14.90

The results for rural high-speed roads are partitioned in Table 19 based on whether the road was an interstate or some other road with a speed limit of 55 mph or more. Although no results are statistically significant, it is noteworthy that involvements in multivehicle

23

crashes are opposing – a 13 percent increase on interstates, compared to a 6 percent reduction on other roads.

Table 19: Number of crashes and reduction for ABS-equipped tractors on roads that are rural and high-speed according to whether the road is an interstate or not, FARS data

The preceding tables indicate that fatal-crash involvements can be classed according to four types of roads and localities: (1) all roads that are not high-speed; (2) roads that are high speed in urban localities; (3) interstate highways in rural localities; (4) other high-speed roads in rural localities. Below are shown the crash reductions for all types of response group crashes on these four road-type/locality combinations. The column for ALL is repeated for convenient comparison.

Table 20: Crash reductions for all crash mechanisms by locality and road type, FARS

	ALL	(1)	(2)	(3)	(4)
All response group	-3%	24%	-3%	-9%	3%
All single vehicle	-5%	21%	-2%	-5%	-4%
Run-off-road					
Overturn	22%	27%	21%	32%	17%
Other	-16%	14%	10%	-5%	-10%
Ped/Bike/Animal	-11%	19%	-19%	-44%	-15%
Jackknife	11%	40%	11%	19%	10%
Other	12%	34%	-7%	26%	-13%
All multi-vehicle response crashes	-1%	26%	-4%	-13%	7%
Impacting in two-vehicle front-to-rear	-43%	35%	-6%	-30%	-33%
At-fault in other multi-vehicle collisions	8%	26%	-4%	-4%	13%

Items shaded in light gray have less than 50 crashes combined for the ABS-no and ABS-yes groups. The percent reduction is presented for informational purposes, but these calculations should be considered less reliable.

(1) all not high speed
(2) not rural - high speed
(3) rural - true interstate
(4) rural - other high speed

The reduction in response group crashes on all non-high-speed roads (1) is much higher than on the other three road types and the only one of the four that is significantly different from zero. The 43 percent increase in rear-impacting another vehicle (ALL) is questionable because it is more negative than on any of the individual road types – e.g., the arithmetic mean of the four estimates is a 9 percent increase. Adjustments will be made that take these factors into account when arriving at a final aggregate estimate.

Revisiting State data based on road and locality

Table 21 presents the results from the State data in the terms of locality and road-speed. The results from FARS are shown for comparison. Results in bold are statistically significant – **black** for reductions; red for increases. Georgia is not included because the data that identifies a crash as inside or outside city limits is unavailable after 1997. Ohio is not included because the explicit rural-versus-urban data element is unavailable after 1999, and it is not clear how to infer locality from what is available from 2000 onward.

As with FARS, the rural and urban results on non-high-speed roads have been combined because the results are similar. Unlike FARS, the high-speed roads have not been further divided into interstate and non-interstate because some combinations are not sufficiently numerous to draw conclusions – the combined results from FARS are shown for comparison, though as Table 19 makes clear, FARS results are best divided based on whether the road was an interstate or some other road with a high speed limit.

Table 21: Reductions for all crash mechanisms on non-high-speed roads, FARS and five States

	FARS	FL	IN	MO	NC	WI
All non-high-speed roads						
All response group	23.7%	19.0%	0.0%	3.7%	6.9%	0.1%
All single vehicle	18.1%	38.0%	0.4%	-9.6%	3.1%	0.5%
All multi-vehicle response crashes	24.8%	6.4%	-0.2%	11.6%	8.7%	3.5%
Rural & high-speed road						
All response group	-10.6%	9.2%	8.7%	16.3%	20.8%	16.6%
All single vehicle	-16.4%	29.1%	5.3%	16.7%	30.6%	18.6%
All multi-vehicle response crashes	-6.8%	-3.9%	11.6%	15.9%	13.7%	13.9%
Non-rural & high-speed road						
All response group	-2.1%	18.5%	27.6%	19.4%	29.0%	19.4%
All single vehicle	-1.9%	33.3%	39.5%	25.9%	31.8%	27.2%
All multi-vehicle response crashes	-1.3%	14.1%	24.2%	17.8%	28.4%	16.5%

There is less variability by road type in the State data than in FARS. The estimates for All response group show a reduction on all three categories of road type in each state. At the next level of All single vehicle and All multivehicle response crashes, the estimates are nearly all positive.

Consideration of other influential conditions for fatal crashes

The type of road where a crash occurred was shown in the previous sections to be influential. Several other conditions are considered that might also be accounted for in arriving at a final estimate. Table 22 shows the percentage of control group crashes, single vehicle response group crashes, and multivehicle response group crashes that occurred under certain conditions. The conditions are defined from FARS as follows:
- Daylight (LGT_COND = 1)
- Level road (PROFILE = 1)
- Not wet (SUR_COND, previously discussed on page 21).
- Driver under 50 ($15 \leq$ DRIVERAGE ≤ 49)
- Straight road (ALIGNMNT = 1)

Table 22: Control and response group crashes for several conditions

	% of Control group		% of single vehicle response		% of multi vehicle response	
	ABS		ABS		ABS	
	NO	YES	NO	YES	NO	YES
Daylight	64%	59%	57%	47%	68%	61%
Level road	73%	72%	64%	65%	70%	69%
Not wet	82%	79%	87%	86%	81%	78%
Driver under 50	69%	67%	65%	62%	72%	66%
Straight road	85%	85%	66%	70%	84%	86%

Only the daylight condition shows varying percentages of crash involvements for the control group and the two response group categories, according to the ABS equipment. The daylight condition is further investigated according to road type.

26

The interaction between road type and light condition is considered first because it could be that the lighting condition is not important on some road types. For eight combinations of road type and ambient light, the 2×2 table of crash involvements was constructed. The odds ratios from these eight 2×2 tables are shown below.

Not high-speed road	Not daylight	0.72
Not high-speed road	Daylight	0.80
Non-rural high-speed road	Not daylight	**1.16**
Non-rural high-speed road	Daylight	**0.91**
Rural interstate	Not daylight	1.10
Rural interstate	Daylight	1.07
Rural other high-speed road	Not daylight	**1.13**
Rural other high-speed road	Daylight	**0.88**

The lighting condition appears influential on two of the four road types – non-rural high-speed roads and rural high-speed non-interstates (highlighted in **bold**).

Arriving at a final estimate for fatal crashes

The final estimate of the effectiveness of tractor ABS in fatal crashes takes into consideration that type of road and locality, as well as lighting condition, are influential and should be controlled for. Six categories are used:

(1) not high-speed roads under all light conditions
(2) rural interstates under all light conditions
(3) non-rural high-speed roads in daylight
(4) non-rural high-speed roads in non-daylight
(5) rural high-speed non-interstates in daylight
(6) rural high-speed non-interstates in non-daylight

This stage of analysis continues to classify crashes according to the mechanisms that have been prevalent throughout the report. The following steps were devised to weight the estimates and provide a statistical significance test:

- compute the *log odds ratio* for each category
- compute the *Variance(logOR)* as $(1/n_{11} + 1/n_{12} + 1/n_{21} + 1/n_{22})$ for a 2×2 table
- construct a weighted *log OR* for each category using the number of ABS-equipped control group crashes as the weight
- construct a weighted *Variance(logOR)* using this same weight, except that the weight must be squared; the *Standard Error(logOR)* is the square root
- place a 95 percent confidence interval around the weight mean of the log odds ratio –*weighted mean(logOR)* ± 1.96 × *Standard error(logOR)*
- convert the weighted log odds ratio and its confidence interval into a weighted odds ratio by raising to the exponential "*e*" value
- subtract the odds ratio and its confidence bounds from one to yield an effectiveness with confidence bounds

Table 23 shows the results of the algorithm above that weights the six conditions based on the control group size of the ABS-equipped tractors.

Table 23: Final weighted estimate of tractor ABS effectiveness from FARS

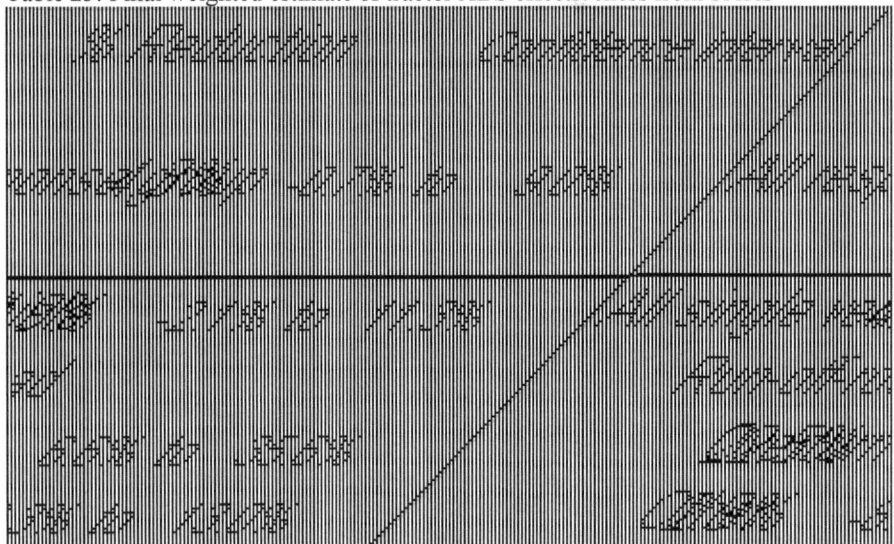

This result can be contrasted with that of Table 15 where no adjustment was made for control group exposure. Table 23 is a more realistic assessment of the driving patterns of ABS-equipped tractors. The difference between the reduction in response group crashes is 6 percent (4% reduction here versus 2% increase previously). Although neither is statistically significant compared to zero, the practical significance of identifying a reduction is favorable.

Table 24 applies the same weighting method applied to the State data, with a starting point being those States included in Table 21. For all response group crashes, the estimated reduction for tractors equipped with ABS is very close for the weighted and unweighted estimates – the difference is at most a little more than one percent. (Slight differences between the unweighted estimates in Table 24 and those in Table 11 are due to exclusion here of a small number of crashes with unknown road type or lighting condition.)

Table 24: Weighted and Unweighted effectiveness for five States

		WEIGHTED				UNWEIGHTED			
		% Reduction	Confidence interval			% Reduction	Confidence interval		
FL	All response group	15.3%	11.1%	to	19.4%	14.6%	10.3%	to	18.6%
	All single vehicle	37.9%	33.2%	to	42.4%	35.8%	31.0%	to	40.2%
	All multi-vehicle response crashes	4.7%	-0.8%	to	9.9%	5.0%	-0.4%	to	10.1%
IN	All response group	10.6%	5.1%	to	15.8%	9.4%	3.9%	to	14.5%
	All single vehicle	16.7%	9.5%	to	23.3%	8.8%	1.7%	to	15.5%
	All multi-vehicle response crashes	8.5%	2.2%	to	14.4%	10.1%	4.2%	to	15.7%
MO	All response group	14.2%	9.3%	to	18.9%	13.0%	8.1%	to	17.6%
	All single vehicle	16.8%	10.1%	to	22.9%	10.0%	3.5%	to	16.0%
	All multi-vehicle response crashes	14.3%	8.8%	to	19.5%	14.9%	9.6%	to	19.9%
NC	All response group	17.2%	11.9%	to	22.1%	17.0%	11.8%	to	21.9%
	All single vehicle	23.9%	16.8%	to	30.4%	23.4%	16.9%	to	29.4%
	All multi-vehicle response crashes	13.7%	7.6%	to	19.4%	13.5%	7.5%	to	19.2%
WI	All response group	11.2%	7.3%	to	15.0%	10.3%	6.4%	to	14.0%
	All single vehicle	16.6%	11.7%	to	21.3%	11.8%	7.2%	to	16.3%
	All multi-vehicle response crashes	7.8%	2.9%	to	12.4%	7.8%	3.0%	to	12.4%

Assessing the influence of vehicle age

The age of the tractor-trailer at the time of the crash is potentially a confounding influence, and an assessment of the effect of vehicle age is typically included in NHTSA regulatory analyses. In a given calendar year of data, the tractors equipped with ABS will be a minimum of two years newer than the vehicles not equipped with ABS. If the likelihood of crash involvement varies in itself according to the age of the vehicle, the results presented thus far could be biased.

One method for assessing vehicle age is to analyze separately each model year, from 1980 to 2007. The graphic below displays the odds (not the odds ratio) of response group crash involvements relative to control group crash involvements (not weighted as in Table 24), in Florida. There is clearly a downward trend in odds for the later model year tractors. Viewed against the ABS effective date, it is reasonable that this trend began around the same time as the ABS mandate in mid-1997. From model year 1995 onward, the trend is nearly monotonically decreasing and always below the cumulative odds from 1980 to 1996. If the downward trend represents improved effectiveness of the ABS, then this result is favorable. On the other hand, if this is evidence of other non-ABS-related innovations, it may be inaccurate to include the latest model years in the calculation of ABS effectiveness. Data from Missouri and Georgia display a similar pattern.

Figure 1: Odds of response-group crash to control-group crash based on tractor model year, Florida State data

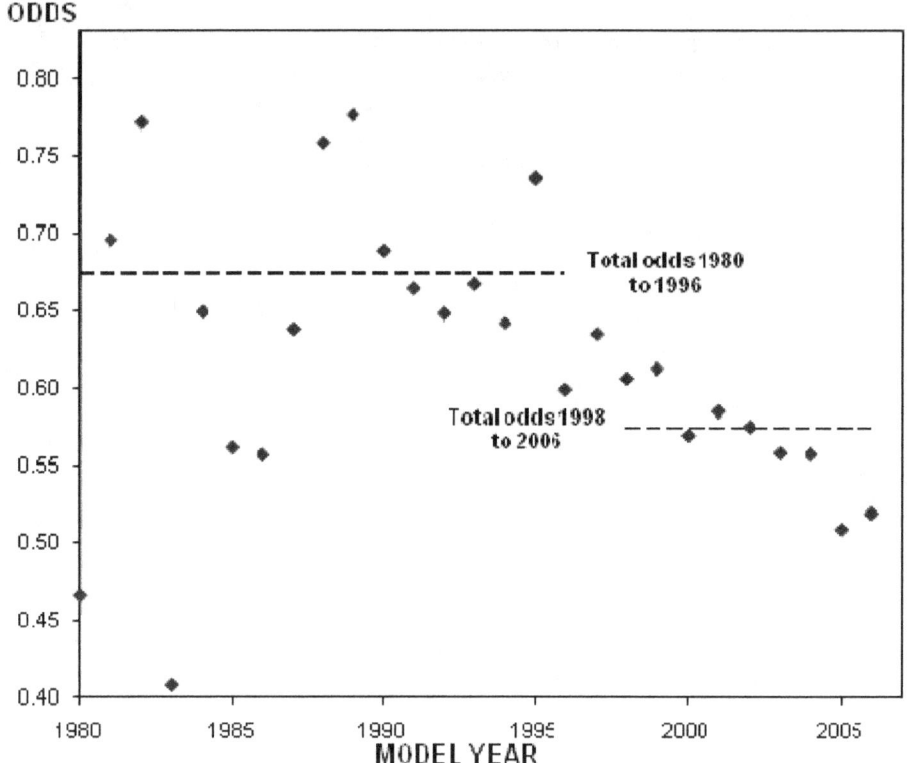

The data from North Carolina has a different pattern, shown below. Here, there is more of a continual downward trend from around model year 1987, though it levels out in the ABS-equipped region beyond 1997. For such a trend, unfortunately, it seems unlikely that 1997 would produce a natural breaking point whereby the ABS mandate could be considered the cause of the decrease. Data from Ohio has a similar trend.

Figure 2: Odds of response-group crash to control-group crash based on tractor model year, North Carolina State data

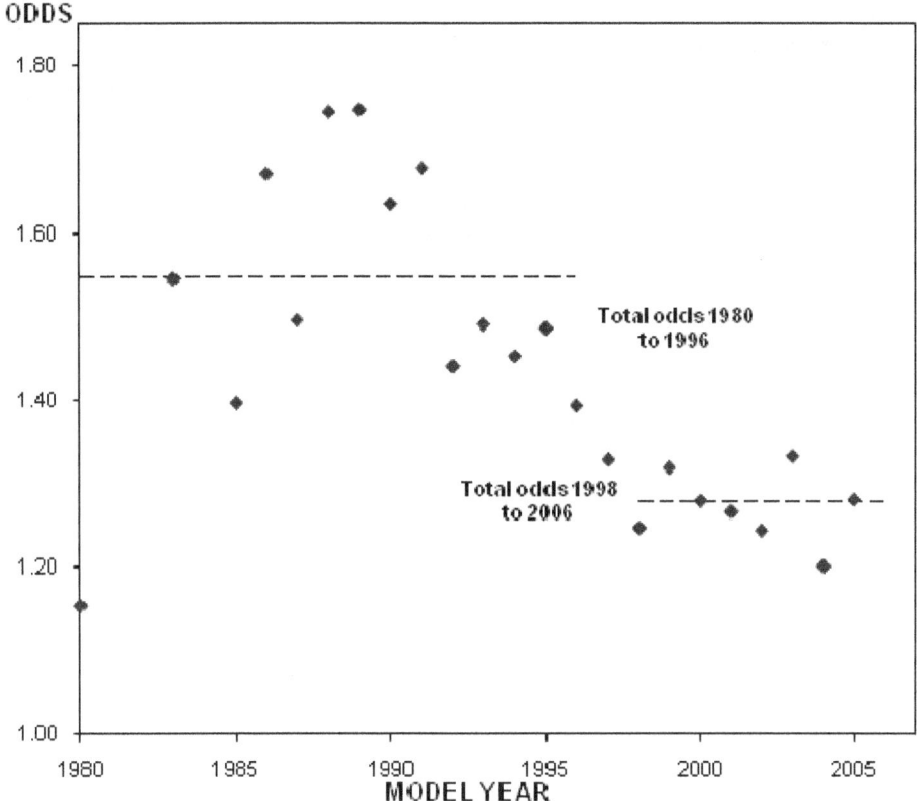

A more ideal pattern is shown below for the data from Wisconsin. The individual model year odds are more nearly scattered around their respective overall odds for the two ABS-equipment regions. Although there is a deal of overlap, there is no apparent downward trend. Data from Indiana is similar.

Figure 3: Odds of response-group crash to control-group crash based on tractor model year, Wisconsin State data

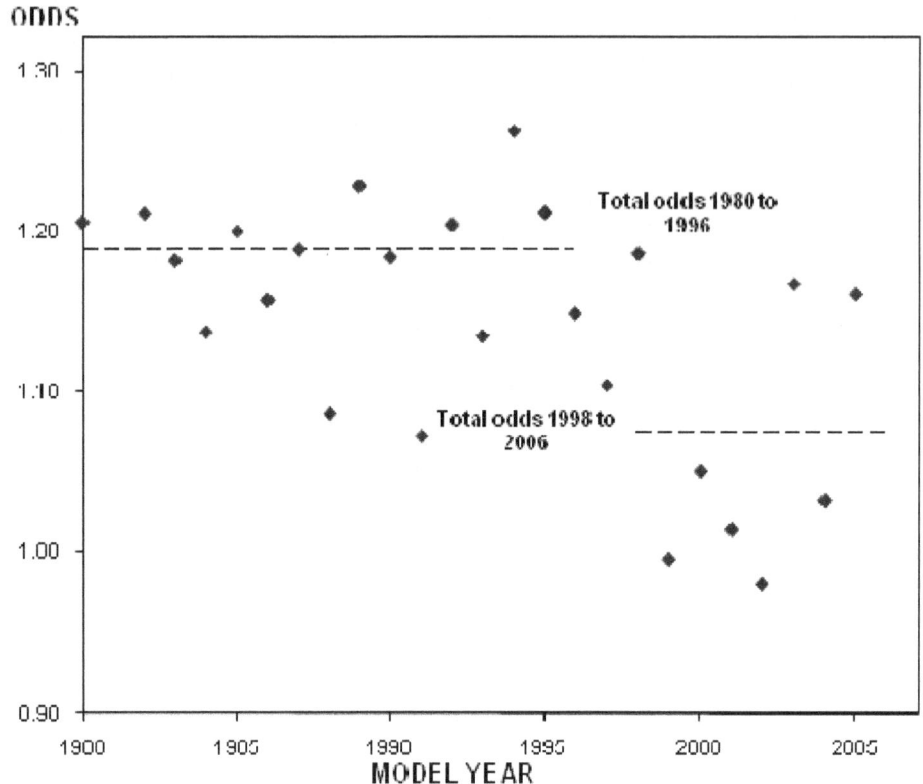

In sum, data from two States (Indiana and Wisconsin) exhibits an approximately ideal relationship between odds and model year. Three States exhibit a downward trend approximately corresponding to the ABS mandate (Florida, Missouri, Georgia). Two States exhibit a trend that does not imply that the ABS mandate in 1997 is the cause of the decreasing odds (North Carolina, Ohio). The comparisons at the level of model year sift the data rather finely, and no statistical significance test is provided on these results.

It appears that vehicle age (using model year as a proxy) may have some effect on the results. As a second step, the data for each State was restricted to vehicle ages that were represented by both ABS-equipped and not-ABS-equipped tractors. In Florida, for example, the minimum age for a non-ABS-equipped tractor is 3 years – a model year 1996 tractor involved in a crash in calendar year 1998, adding one to adjust for the fact that model years begin several months in advance of calendar years. This creates a lower-bound on the ages at 3 years. The upper-bound is based on the earliest ABS-equipped tractors – a model year 1998 tractor involved in a crash in 2007 is up to 10 years old.

The subset of vehicle ages varies by State, according to the available calendar years of data. Below are listed the vehicle ages included for each State. Vehicle ownership and the types of routes driven could vary by age – e.g., long-distance runs for fleet-owned vehicles when they are new, transferring to owner-operators later and more short-distance, local routes.

Florida	$3 \leq \text{age} \leq 10$	
Georgia	$3 \leq \text{age} \leq 9$	
Indiana	$7 \leq \text{age} \leq 8$	
Missouri	$7 \leq \text{age} \leq 9$	
North Carolina	$8 \leq \text{age} \leq 9$	
Ohio	$5 \leq \text{age} \leq 9$	
Wisconsin	$3 \leq \text{age} \leq 9$	

Below are the results by State for the subset of vehicles of compatible ages (labeled **AGE**) compared to the overall unweighted results (labeled **ALL**). The differences at right (**AGE** *minus* **ALL**) show that when this measure of vehicle age is accounted for, the effectiveness is lower in six of the seven States. Three States (FL, GA, OH, shown in **bold**) have effectiveness that is statistically significant *from zero* at the 0.05 level. The median of the seven age-controlled effectiveness calculations is 6.3 percent (that of Florida, shown at the bottom). This compares with a median of 12.7 percent for the complete datasets. Some effect due to vehicle age is likely present. A recent NHTSA analysis of repair and maintenance expenses for heavy vehicle ABS showed the system to be quite reliable – there were, on average, 21 repairs per 100 tractors in a survey of fleet repair receipts over approximately a two-year period (DOT HS 811 109, March 2009) – thus, age effects are more likely due to differences in vehicle usage.

Table 25: Overall response-group reduction for age-restricted datasets compared to complete datasets

	(AGE)		(ALL)		(AGE) minus (ALL)
	Reduction	$\chi 2$	Reduction	$\chi 2$	
FL	6.3%	3.74	**13.9%**	37.70	-7.5%
GA	**7.4%**	17.50	**11.2%**	78.66	-3.8%
IN	10.2%	1.96	**9.7%**	11.74	0.6%
MO	5.2%	0.82	**12.9%**	25.00	-7.7%
NC	4.8%	0.33	**16.7%**	34.89	-11.9%
OH	**7.2%**	8.57	**12.7%**	74.51	-5.5%
WI	4.8%	3.03	**9.4%**	21.37	-4.6%
Median	6.3%		12.7%		

As suggested in Table 10, ABS may be more effective on wet roads. In Table 26, the overall reduction in response group crashes is shown for the **AGE** and **ALL** datasets. Two factors are notable – (1) ABS is more effective on wet roads compared to all road conditions; (2) the necessity of controlling for age appears less important. On the other hand, these calculations are less precise because only around 20 percent of crashes are classified as being on wet roads.

Table 26: Overall response-group reduction on wet roads for age-restricted datasets compared to complete datasets

	(AGE)		(ALL)		(AGE) minus (ALL)
	Reduction	χ2	Reduction	χ2	
FL	**20.8%**	11.08	**23.6%**	27.75	-2.9%
GA	**17.3%**	30.56	**18.8%**	67.58	-1.5%
IN	-22.4%	2.80	**10.9%**	6.52	-33.3%
MO	11.9%	1.74	5.1%	1.37	6.8%
NC	**38.8%**	11.58	**17.2%**	13.28	21.6%
OH	4.5%	1.42	**7.2%**	9.29	-2.6%
WI	**15.0%**	14.14	**16.3%**	28.61	-1.3%
Median	15.0%		16.3%		

A simultaneous assessment of the effects of vehicle age and ABS equipment was performed using a logistic regression model. The 2×2 table was modeled with control versus response group crash involvement as the dependent variable and ABS equipment as the independent variable. Vehicle age was incorporated as a covariate. Both the age-restricted (**AGE** in the above discussion) and the full dataset (**ALL**) were analyzed. The results are summarized in the following table and discussed thereafter.

(**ALL** dataset)

	OR on ABS	p-value on ABS	p-value on vehicle age
FL	0.985	0.6630	< 0.0001
GA	0.914	< 0.0001	0.0253
IN	0.961	0.4126	0.1167
MO	0.947	0.2301	0.0206
NC	0.905	0.0594	0.0507
OH	0.955	0.0583	< 0.0001
WI	0.935	0.0108	0.1027

Median of all 7 OR   0.955 (OH)
Effectiveness for median OR $1 - 0.955 = 0.045 = 4.5\%$

(**AGE** dataset)

	OR on ABS	p-value on ABS	p-value on vehicle age
FL	1.044	0.2869	< 0.0001
GA	0.933	0.0007	0.2420
IN	0.855	0.0569	0.0884
MO	0.943	0.4004	0.3806
NC	0.933	0.4799	0.7916
OH	0.948	0.0702	0.1666
WI	0.934	0.0263	0.2621

Median of all 7 OR   0.934 (WI)
Effectiveness for median OR $1 - 0.934 = 0.066 = 6.6\%$

The **AGE** dataset pretty well controls for the effect of age in its design – only one of the seven States (Florida) has a significant effect for the covariate of vehicle age.

34

Finally, the **AGE** dataset is further analyzed in Table 27 according to the relevant crash mechanisms. This can be compared to Table 11 for all ages of tractors. All States have reductions in All response group crashes, but only three are statistically significant (FL, GA, OH). One number that stands out is a large increase in the occurrence of rear-ending another vehicle in Florida (46% increase) – no other State has a statistically significant result. Most importantly, the results are relatively consistent for the other crash types in the seven States. Few results are statistically significant with this restricted dataset. At the finest level of detail, there are 49 comparisons with only two being statistically significant increases – it is reasonable that such occurrence could be chance variation.

The overall reduction in crashes, on the bottom row of Table 27, is lower than that predicted in the Final Economic Assessment[1] (8.86%, based on re-analysis of a 1984 German study of 182 crash involvements). This early assessment was based on expert judgment as to whether a crash would have been prevented if the heavy vehicle had been ABS-equipped. It was not possible to perform a statistical analysis based on response and control group crashes at that time when ABS was rare and non-mandated for what few vehicles may have had it installed.

Table 27: Reduction in all crash mechanisms, age-restricted State data

	FL	GA	IN	MO	NC	OH	WI
All response group	8%	8%	10%	3%	6%	7%	5%
All single vehicle	39%	21%	5%	7%	12%	5%	9%
Run-off-road							
Overturn	18%	23%	67%	19%	18%	-3%	-3%
Other	33%	22%	-48%	4%	21%	6%	10%
Animal/Ped/Bike	-25%	12%	30%	-16%	-29%	-20%	12%
Jackknife	76%	72%	22%	73%	78%	42%	49%
Other	49%	12%	-1%	6%	-14%	11%	6%
All multi-vehicle response crashes	-14%	3%	13%	0%	2%	9%	2%
Impacting in two-vehicle front-to-rear	-46%	1%	7%	-3%	20%	-1%	1%
At-fault in other multi-vehicle collisions	5%	6%	15%	2%	-7%	13%	2%
Overall reduction in crashes	3%	4%	6%	2%	3%	4%	3%

In FARS, the range of overlapping vehicle ages is 3 to 11 years – this range is wider than for the State data because FARS has maintained consistent variable definitions over a longer period. The list below shows the reduction in response-group crashes across the range 3 years to 11 years. The calculations were also performed at each individual age – based on a 2×2 table restricted to vehicles of the age listed. Unlike with the State data, the results do not vary greatly according to age of the tractor.

35

	Reduction	χ^2
Full range of ages (Table 15)	-2%	0.97
Restricted to $3 \le age \le 11$	-4%	1.70
age = 3 only	-5%	0.14
age = 4 only	-7%	0.45
age = 5 only	-25%	5.91
age = 6 only	-5%	0.28
age = 7 only	-9%	0.84
age = 8 only	+4%	0.22
age = 9 only	-5%	0.17
age = 10 only	+9%	0.45
age = 11 only	+1%	0.00

As with the State data, a logistic regression model was fit using ABS equipment with vehicle age as a covariate. Below are the summary statistics for two models. First is the model without restriction in the vehicle ages. Second is a restricted dataset where only the overlapping tractor ages are included. As implied by the list above, vehicle age makes little difference, being a non-significant effect in both datasets, and the odds ratio estimates differ by only 1.8 percentage points. The weighted estimate based on road type and lighting condition remains the best estimate of effectiveness in fatal crashes.

	OR on ABS	p-value on ABS	p-value on vehicle age
All ages	1.037	0.2885	0.4555
$3 \le age \le 11$	1.055	0.1272	0.4156

Limiting Factors

Two factors limit the perceived effectiveness – the extent of voluntary ABS installations prior to the mandate and the number of vehicles with non-functioning or disabled ABS. To illustrate the potential influence of these factors, a hypothetical analysis below shows an effectiveness calculation of 10 percent. From the State data, the estimate based on all tractors was 12.7 percent and 6.6 percent based on the dataset restricted to tractors of overlapping ages – the value 10 percent is convenient for illustrative purposes as a round number that falls between these two estimates.

	Control	Response	Odds for this row
1996 & older ALL	1000	1000	1.00
1998 & newer ALL	1000	900	0.90

Odds ratio	$(900 / 1000) \div (1000 / 1000) = 0.90$
Effectiveness	$1 - 0.90 = 0.10 = 10\%$

The potential influence of pre-mandate tractors with ABS is illustrated below, based on 10 percent of 1996 model year and older vehicles having ABS. This value is estimated from a 2009 NHTSA report about the costs to maintain and repair ABS in tractors[3] – an expense of 10¢ per month is reported for tractors with a model year 1996 and earlier (an average that includes primarily non-ABS-equipped tractors), compared to 85¢ per month for tractors with ABS mandated by FMVSS 121 ($10 \div 85 = 11.8$ percent; round down to 10% for simplicity). It is assumed that these would exhibit the same reduction in odds as the 1998 and newer vehicles. That is, the odds of 0.90 for the 1998 and newer row is applied to the response group crashes from the 1996 and older row (10% of 1000 is 100). The control group is adjusted similarly ($100 \div 0.90 = 111$). Then, the remaining 1996 and older tractors without ABS are 889 in the control group ($1000 - 111$) and 900 in the response group ($1000 - 100 = 900$). The adjusted effectiveness calculation is 11 percent.

	Control	Response	Odds for this row
1996 & older 90% without ABS	889	900	1.01
1996 & older 10% with ABS	111	100	0.90
1998 & newer ALL	1000	900	0.90
Sum of two above with ABS	1111	1000	0.90

Odds ratio	$(1000 / 1111) \div (900 / 889) = 0.890$
Effectiveness	$1 - 0.890 = 11\%$

On the other hand, some tractors with ABS may have the system disabled or it may not be functioning properly. A study of in-service tractor-trailers was conducted by the

---

[3] Allen, K. (2009). *An In-service Analysis of Maintenance and Repair Expenses for the Antilock Brake System and Underride Guard for Tractors and Trailers*. DOT HS 811 109. Washington, DC: National Highway Traffic Safety Administration. http://www-nrd.nhtsa.dot.gov/Pubs/811109.pdf

Federal Motor Carrier Safety Administration in 2004.[4] Inspections were conducted at weigh stations in four States, and 4 percent of tractors produced after March 1, 1997, were found to have an ABS system fault. The percentage of crash-involved tractors with an ABS fault might be different than the population of in-service vehicles. The effectiveness can be adjusted in a manner similar to the example above, where now there are 4 percent of the 1998 and newer tractors in the response group that are considered to not have ABS. Because 4 percent is small, the change in effectiveness is negligible. The methodology for re-estimating the effectiveness could be applied for some other proportion.

	Control	Response	Odds for this row
1996 & older ALL	1000	1000	1.00
1998 & newer 4% inactive ABS	36	36	1.00
1998 & newer 96% functioning ABS	964	864	0.896
Sum of top two rows	1036	1036	1.00

Odds ratio	$(864 / 964) \div (1036 / 1036) = 0.896$
Effectiveness	$1 - 0.896 = 10\%$

[4] Shaffer, S. J. (2005). *Assessment of ABS Malfunction Indicator Lamp Status – A Snapshot of In-Service Vehicles.* DOT-FMCSA-MCP/PSV-05-003. Washington, DC: Federal Motor Carrier Safety Administration.

www.ingramcontent.com/pod-product-compliance
Lightning Source LLC
Chambersburg PA
CBHW081358170526

45166CB00010B/3135